Caio Braga Ferreira
Francisco Guimarães
Francisco Edu De Andrade

Experiments in Agroecology: An Approach with Passion Fruit and Lettuce

Caio Braga Ferreira
Francisco Guimarães
Francisco Edu De Andrade

Experiments in Agroecology: An Approach with Passion Fruit and Lettuce

An agro-ecological analysis of Yellow Passion Fruit cultivars and soil chemical attributes after growing Lettuce

ScienciaScripts

ACKNOWLEDGEMENTS

To my wife Fabiane A. do N. Bristotti, for her efforts, encouragement and understanding over all these years.

To my supervisor Marshell F. A. Ferraz, for his dedication, companionship and trust during the preparation of this work.

To the company represented here by Mr Alexsandro Pereira - the owner who gave me the support and space to carry out this work.

To Univali Universidade do Vale do Itajaí, and all its teaching and administrative staff, for providing me with high-quality study.

To my undergraduate friends, with whom I had the pleasure of sharing a lot of joy and knowledge over the years.

To God for giving me the opportunity to meet and socialise with all these people.
And to all those who contributed in some way to this work.

"You can't manage what you don't measure, you can't measure what you don't define, you can't define what you
don't understand, there is no success in what you don't manage" (William Edwards Deming).

SUMMARY

BRISTOTTI, Elvis. Analysing slag reduction in welding between steel plates in the naval production area using the scientific analytical method. Itajaí, 2016. Scientific and Technological Initiation Work (Undergraduate Degree in Production Engineering) - Centre for Land and Sea Technological Sciences. Univali Universidade do Vale do Itajaí, Itajaí, 2016.

The shipbuilding industry strengthens the economy of many sectors, such as the steel industry, metallurgy, machinery and equipment, information technology and others. As a result, it represents a large slice of the country's domestic product, generating more than 50,000 direct jobs and many thousands more indirectly. With the crisis that has hit the sector, it is essential that there is stricter economic control in order to increase productivity at the lowest manufacturing cost and with the same quality. In shipbuilding, welding is the most widely used of all the processes, as most of the various parts of a vessel are welded together. Welding discontinuity is any type of fault that may appear in a welding job and jeopardise the use and function of the object being welded. There are various types of fault such as: cracks, bites, porosity, lack of fusion and others. The aim of this work was to analyse the reduction of slag in the first weld bead, using the scientific analytical method (where test specimens were made with laboratory modifications of the process tests). The specimens were analysed using ultrasound testing to check for changes in slag formation as the process data changed). By following the parameters stipulated in the methodology of this work, it was possible to see that by changing some of the standard requirements of the welding process, a different result can be obtained, so it can be concluded that one of the parameters presented was successful in reaching the optimum point required by the welding process.

Keywords: Welding, Naval Industry, Slag, Welding discontinuity.

Summary

CHAPTER 1

INTRODUCTION

1.1 Background and justification

The shipbuilding industry strengthens the economy of many sectors, such as the steel industry, metallurgy, machinery and equipment, information technology and others. As a result, it represents a large slice of the country's domestic product, generating more than 50,000 direct jobs and many thousands more indirectly (SALSA, 2009 apud BROGNOLI, 2016).

Data from the National Union of the Naval and Offshore Construction and Repair Industry (Sinaval) shows that the industry has been almost halved in less than two years. At the end of 2014, there were more than 82,000 employees; the total fell to 43,000 in June 2016. With the crisis that has hit the sector, it is essential that there is stricter control on the economic side, seeking to increase productivity at the lowest manufacturing cost and **with the same quality: "Thus, knowing the** cost structure becomes paramount for the success of **companies."** (MARQUES et al. 2011).

In shipbuilding, among all the processes, welding is widely used, as most of the various parts of a vessel go through this process.

"Welding processes have a wide range of applications, including, among others, shipbuilding, civil structures, pressure vessels, pipelines, various types of equipment, hydroelectric power stations, underground and railway materials and **nuclear** components." (HORMEU, 2008).

There is a difference between the terms welding and soldering. Welding is the process by which a joint is made. Welding is the joining zone where solubilisation has taken place.

"Welding is the most important metal joining **process** used **industrially".** (MARQUES et al. 2011). Within this process there are different types of welding, such as: gas welding and cutting, coated electrode welding, TIG welding (**Gas Tungsten Arc Welding - GTAW**), plasma welding and cutting, submerged arc welding, electroslag and electrogas welding and resistance welding.

The importance of welding in industry today shows how this sector has become specialised in the product it supplies, and that it often devotes less effort to other activities that are also part of the production process.

In order to carry out quality welding, attention must be paid to certain parameters: current intensity, voltage and arc length, welding speed, wire length, shielding gases, electrode diameter and torch position and flow rate (QUITES, 2002).

Welding discontinuity is any type of fault that may appear in a welding job and which jeopardises the use and function of the object being welded. There are various types of fault such as cracks, bites, porosity, lack of fusion and slag inclusions (MARQUES et al. 2011).

In this work, the problem studied was the existence of welding slag inside the weld, which led to poor quality and rework. There were several factors that could be influencing the quality of this weld, such as: welding equipment not being adjusted according to the project parameters, cleaning not being carried out properly according to the process and a lack of training on the part of the welding professional.

This study was designed to analyse the reduction of slag in the first weld bead, with the aim of helping the process to be carried out efficiently and without waste. Once a discontinuity is detected, rework is inevitable and this leads to losses such as wasted material and labour time, causing delays in the product delivery schedule.

1.2 Objectives

1.2.1 General Objective

Finding the optimum welding point between ASTM A36 steel plate joints in order to reduce slag formation in the first bead and consequently reduce costs and time spent on reworking the process.

1.2.2 Specific objectives

- Analysing the structure of welding between ASTM A36 steel plates
- Characterise the formation of slag between ASTM A36 steel plates in the first weld bead using ultrasound techniques.
- Verify the best method for applying the second layer of solder.

CHAPTER 2

THEORETICAL BACKGROUND

2.1 The Brazilian shipbuilding industry

According to Santos (2011), when we talk about the Brazilian shipbuilding industry, we can't help but mention Irineu Evangelista de Souza, the forerunner in this field and better known as the Baron and Viscount of Mauá. A visionary entrepreneur in the industrial and commercial sector, he founded the Ponta da Areia Niterói (RJ) shipyards in 1845, giving birth to the Brazilian shipbuilding industry.

However, it wasn't until 1958 that this sector was given special consideration, when the then president of Brazil, Juscelino Kubitschek, stipulated a target plan. Law No. 3.381, which was passed in April 1958, planned to raise funds for the renovation, extension and recovery of the shipbuilding sector.

According to the author, during the governments of Jânio Quadros (1961), João Goulart (1961 - 1964), Ranieri Mazilli's interregnums (in 1961 and 1964) and the Castelo Branco government (1964 - 1967). Brazilian shipbuilding remained simple and regular.

The Brazilian shipbuilding industry has spread throughout the various regions of the country, concentrated in the states of Amazonas, Pará, Ceará, Pernambuco, Sergipe, Bahia, Espírito Santo, Rio de Janeiro, São Paulo, Santa Catarina and Rio Grande do Sul.

In shipbuilding there are a variety of manufacturing processes, and welding is one of the most widely used, as most of the various parts of a vessel go through this process.

"Welding processes have a wide field of application, including, among others, shipbuilding, civil structures, pipelines, various equipment, **hydroelectric power stations, underground and railway materials" (HORMEU, 2008).**

2.2 Importance of welding

Welding is the process of joining two or more parts, ensuring that the joint has continuity of chemical and physical properties (QUITES, 2002).

According to Quites (2002), in order to guarantee continuity, additional material must be added to complete the space between the base materials, and the addition material must dissolve in the base material.

According to Machado (1996) apud Machado (2015), welding is the process that stands out most among the processes of joining materials, mainly due to its wide use and the large volume of activities it involves. Its main area of activity is metals and their alloys, which is due to its great versatility and economy, as well as the excellent mechanical properties of the joints obtained in this way.

Wainer et al (1992), who call welding the process of joining two metal parts using a heat principle, which may or may not use pressure, the result of this process is the weld.

According to Modenesi et al (2011), welding is the method of joining elements based on the creation of chemical bonding forces of a nature equivalent to those involved in the material itself inside the place where the elements to be fused are being joined.

There are a large number of processes that can be defined as welding. Generally speaking, welding is known as a method of joining materials, whether metallic or not. However, in many of the variations that exist today, this process is used to repair parts by overlapping materials in order to recover a worn area or even to form a coating. There are also variants that are used to cut metal parts and, in many respects, are similar to welding processes (MODENESI et al. 2011 apud BROGNOLI, 2016).

According to Quites (2002), welding is a physical process that originates from an exclusively metallurgical method. Its purpose is to unify components using the welding explanation.

Modenesi et al. (2011) also point out that:

> Metal joining methods can be divided into two main categories: those based on the action of macroscopic forces between the parts to be joined and those based on microscopic interatomic and intermolecular forces. In the first case, of which screwing and riveting are examples, the strength of the joint is given by the shear strength of the screw or rivet plus the frictional forces between the surfaces in contact. In the second category, the joint is achieved by bringing the atoms or molecules of the parts to be joined, or of these and an intermediate material added to the joint, close enough to form chemical bonds, particularly metallic and Van der Waals bonds. Examples of this category include brazing, welding and gluing.

2.3 MIG/MAG welding

Gas shielded welding processes began in the 1920s.
However, it wasn't until the 1950s that the combination of CO_2 shielding gas with electrodes containing internal flux (flux-cored wires) was introduced, which led to significant improvements in operating conditions and weld quality. (BARBEDO, 2008 apud MACHADO, 2015

Fortes (2006) defines the electric arc process with tubular wires as the process that results from the joining of metals by an electric arc established between the welding wire and the base metal to be welded. In this process, the protection of the electric arc consists of the internal flux of the wire and the option of adding or not adding shielding gas.

According to Machado (1996), this welding process is one in which the electric arc is formed by the contact between the base metal and the welding wire, from which the internal filling by fusible flux is obtained, which is constantly fed by a coil. The molten pool generated at this stage is covered in slag and gases resulting from the process. Shielding gas can be added, where the gas is released from the torch nozzle.

According to Modenesi et al. (2011), electric arc welding with shielding gas is a process that produces the joining of metals by heating them with an electric arc established between a continuous, consumable bare metal electrode and the workpiece.

The device used in MIG/MAG welding can be semi-automatic or automatic. In the semi-automatic device, the consumable wire is passed automatically through the equipment and the other process interventions are carried out by the welder. However, in the automatic device, the welder has no influence on the process; he only adjusts the device before starting the welding process (JUNIOR; CABRAL, 2008).

MIG/MAG welding has been widely used in the automotive industry, particularly with the use of robots. As it is a semi-automatic process and uses high current densities, its productivity is quite high. In addition, it can also be mechanised relatively simply with the use of torch positioning and displacement devices (MODENESI et al. 2011 apud BROGNOLI, 2016).

Welding with a metal arc and covered by a shielding gas is a technique in which the dissolution that occurs at the edges to be joined and of the filler metal are generated by the influence of a visible electric arc determined by a consumable electrode, constantly fed to the molten pool, and the base metal (QUITES, 2002).

This protection of the arc or the weld region against atmospheric contaminants is done by a gas or a mixture of gases, which can be inert or active. In Brazil, the process is known as MIG when the shielding used is inert or rich in inert gases or MAG when the gas used is active or contains mixtures rich in active gases (MODENESI et al. 2011).

Marques et al. (2011) point out that:

> The MAG process is only used for welding ferrous materials, with CO_2 or mixtures rich in this gas as the shielding gas, while MIG welding can be used for welding both ferrous and non-ferrous materials, such as aluminium, copper, magnesium, nickel and their alloys.

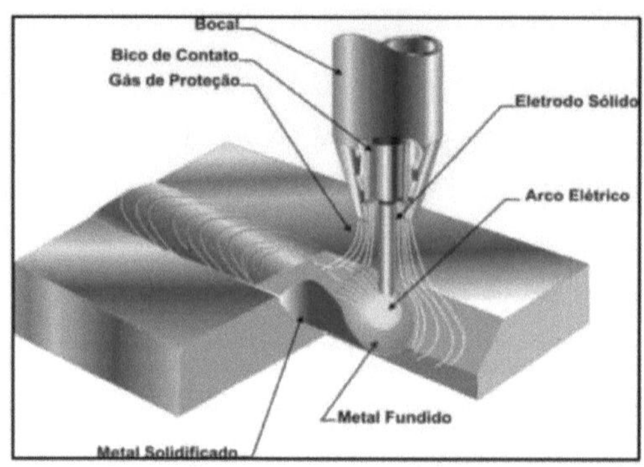

Figure 1 Illustration of the MIG/MAG process Source: Junior; Cabral, 2008

MIG/MAG welding is used on ferrous and non-ferrous materials over a wide range of thicknesses, as shown in Table 1. The diameter of the electrode used is usually between 0.8 mm and 2.4 mm (MODENESI et al. 2011).

Table 1 Procedure for various thicknesses to be welded using the MIG/MAG process

Procedure	Thickness (mm)	
	Minimum	Maximum
Single pass without preparation	1,0	5,0
Single pass with preparation	3,2	10
Multiple passes	3,5	XX

Source: (QUITES, 2002).

The MIG/MAG method is a type of welding used in the manufacture and maintenance of equipment, the restoration of worn parts and the covering of metal surfaces with unique elements (QUITES, 2002).

According to Modenesi et al. (2011), [...] some of the advantages of MIG/MAG welding compared to covered electrode welding are: high deposition rate and high welder occupation factor, great versatility in terms of the type of material and thickness applicable, no welding flux and, consequently, no slag removal and cleaning operations [...].

This type of welding has seen considerable growth in recent years in terms of its use worldwide. This is due to the need to upgrade this process to semi-automatic and mechanised equipment whenever possible, in order to achieve a higher level of productivity during welding (MODENESI et al. 2011).

2.4 Discontinuities in welding

Modenesi et al. (2011) consider a discontinuity to be a break in or defilement of the natural or desired shape of a welded joint.

According to Novais (2010), "it is an interruption in the typical structure of the weld, such as a lack of homogeneity in its mechanical, metallurgical or physical characteristics. A **discontinuity is not necessarily a defect. "**

Fortes (2005) points out some of the defects that can appear in weld metal, which are shown in figure 2.

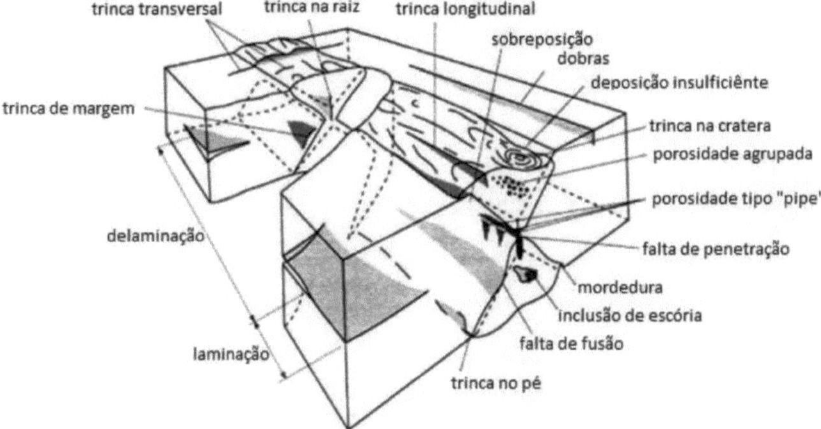

Figure 2 The different discontinuities in an X-bevelled butt joint can be seen in the schematic drawing. Source: Infosolda, 2016[1]

2.3.1 Cracks

Modenesi et al. (2011) considers a crack to be the most serious discontinuity possible in a weld. A crack can be the consequence of the material's inability to resist the localised stresses exerted by the welding process. With the application of localised heat, tensile stresses appear near the welded region, and the fragility of the weld, together with microscopic changes during the process and the presence of certain elements such as hydrogen, can result in cracks.

The discontinuity can be generated in various ways. There are some devices that make it easier to form cracks in welds, which will be described in brief below (MODENESI et al. 2011).

As already mentioned, they have a high tensile capacity, contributing to the appearance of a weak rupture in the welded support (NOVAIS, 2010).

2.3.2 Hot cracks

Hot cracking, also known as solidification cracking, is a very common form of cracking and

[1] The different discontinuities in an X-bevelled butt joint can be seen in a schematic drawing. Infosolda. Available at: <http://www.infosolda.com.br/biblioteca-digital/28-biblioteca-digital/livros-senai/metalurgia/125-descontinuidades.html> Accessed on: Sep. 2016.

can occur in other manufacturing processes, specifically casting. It is associated with the formation of microscopic films of separated liquid material that cannot withstand the stresses suffered during the solidification process of the material (MODENESI et al. 2011).

Another very important factor that can influence the formation of cracks during solidification is the shape of the strand, because depending on the amount of liquid metal feed that the strand region receives, it can influence the stresses that act on these regions (MODENESI et al. 2011). Figure 3 shows the effect of bead shape on the formation of this type of crack.

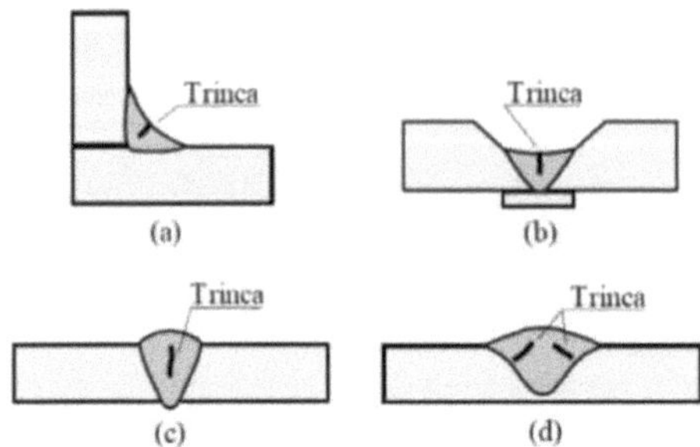

Figure 3 Hot crack formation favoured by bead shape in: (a) a concave fillet weld and (b) concave root pass in butt joint, (c) butt weld with high penetration/width ratio and (d) in a bell-shaped bead.
Source: MODENESI et al. 2011.

2.4. 3Cold cracks

Cold cracking is also known as hydrogen cracking, this type of crack is very common in carbon, low and medium alloy steels, especially those that are hardened during welding (MODENESI et al. 2011).

According to Fortes (2005), hydrogen cracking occurs when hydrogen comes into contact with the weld pool, either through humidity or effective compounds in the flux or on the surface of the base or filler metal. In the area of the weld where the bead has already solidified, the retained hydrogen can be seen giving rise to cracks, which can be seen in Figure 4.

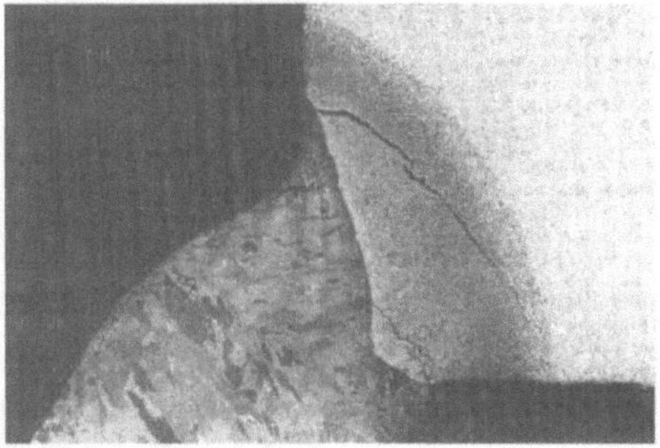

Figure 4 Hydrogen cracks.

Source: Fortes, 2005

2.4. 4Porosity

According to Novais (2010), porosity is a vague environment produced by the enclosure of gas while solidification takes place.

Porosity can occur in three ways. Firstly, as a result of chemical reactions in the weld pool, i.e. if a steel weld pool is inadequately deoxidised, the iron oxides can react with the carbon present to release carbon monoxide (CO). Porosity can occur at the start of the weld bead in manual welding with a coated electrode because at this point the protection is not fully effective. Secondly, by the expulsion of solution gas as the weld solidifies, as happens when welding aluminium alloys when hydrogen from moisture is absorbed by the puddle and later released. Thirdly, by the trapping of gases at the base of turbulent fusion pools when welding with shielding gas, or the evolved gas when welding the other side of a T-joint on a sheet with bottom paint (FORTES, 2005).

According to Fortes (2005), most of this discontinuity is easy to resolve and can be avoided if the cleaning parameters are followed correctly. Porosity is not a defect that causes excessive damage to mechanical properties, except when it appears on the surface. If it does, it can lead to cuts that cause premature fatigue failure. Figure 5 shows the appearance of some discontinuities such as porosity on the inside of the weld.

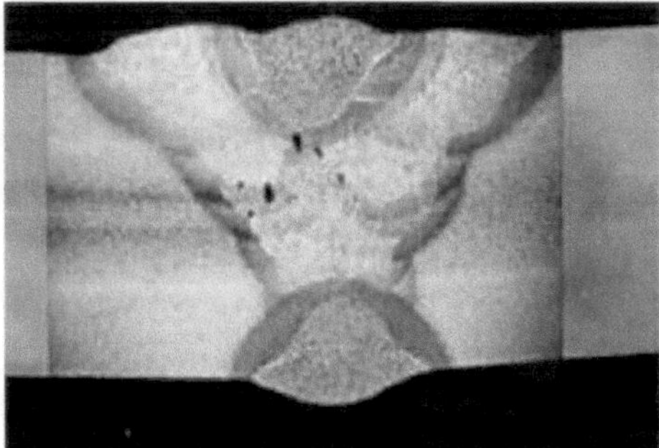

Figure 5 Micrograph showing porosity Source: NOVAIS, 2010.

At the moment when the weld bead is solidifying in the weld pool, some gas bubbles may appear, this is a consequence of impurities in the process, either due to humidity, some type of oil or even rust, they remain trapped and can damage the welded part or not. (MODENESI et al. 2011). Figure 6 shows surface porosity.

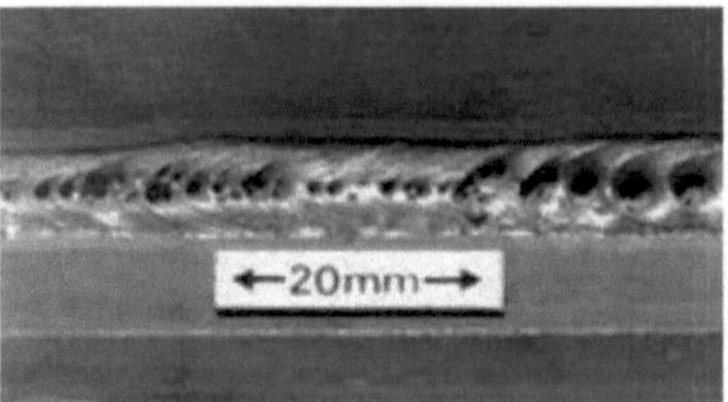

Figure 6 Surface porosity

Source: Novais, 2010.

Table 1 below shows the main causes and possible solutions for porosity.

Table 1 The main causes and possible solutions for porosity

Causes	Possible solutions
Small shielding gas flow	Increase the flow of shielding gas
Obstruction of the gas outlet due to the amount of splash in the nozzle	Clean the nozzle regularly
Turbulence due to excessive shielding gas flow	Reduce the flow of shielding gas
Wind at the welding site	Adequately protect the welding area
Oxidised or dirty electrode wire	Use only clean, dry wire electrodes
High welding speed causing insufficient protection	Reduce welding speed
Torch too far away from the base metal	Bring the torch close to the base metal and, at the end of the bead, hold it over the molten pool until it solidifies.
Excessive torch travel angle	Decrease the angle appropriately
Contaminated base metal (oxidation, grease, oil or paint)	Improve the cleanliness of the base metal
Improper voltage and current setting (high voltage / low current)	Adjust welding voltage and current appropriately
Contamination of the protective gas	Periodically check the sealing of the protective gas system and the quality of the gas used
Wire-electrode class unsuitable for the type of shielding gas used	Use wire-electrode suitable for the type of gas used in welding
Base metal with S content > 0.05% or with a high number of inclusions	Use base metal with an S content of <0.05% and with few inclusions

Source: NOVAIS, 2010 apud FILHO, 2012

2.4.5 Inclusion of slag or other inclusions

In the process that takes place in the foundry and immediately after solidification of the melting pool, a large number of effects are processed. These effects can produce some elements which, if trapped in the solidified metal, give rise to inclusions (NOVAIS, 2010).

Flux is a component used in the welding process to help the metal adhere, and it has some particles that can become trapped in the weld pool, thus generating inclusions in the weld bead (FORTES, 2005).

Figure 7 shows an example of an X-ray showing slag inclusions:

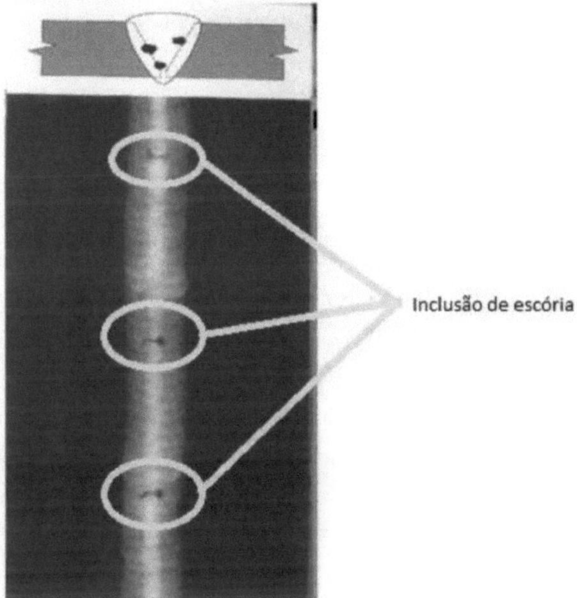

Figure 7 X-ray showing slag inclusions Source: Novais, 2010.

According to Fortes (2005), inclusions are more likely to occur in the course of the process during subsequent passes or even in the weld metal and the base metal chamfer. Inadequate cleaning of these parts can be considered the most common cause of this type of discontinuity.

As with porosity, detached inclusions do not cause disadvantages to mechanical properties, so inclusions aligned in certain specific locations, such as in the direction transverse to the tension used, contribute to the appearance of fracture (FORTES, 2005).

When carrying out some types of welding, fluxes are used that can form slag in the middle of the weld pool that can separate from the liquid metal. There are other reactions that take place in the molten pool, where insoluble products are formed in the liquid metal that are capable of separating and likewise generating slag. Some of the slag created in the weld pool tends to be trapped between the weld seams or even in the base metal, as shown in Figure 8 (MODENESI et al. 2011).

Figure 8 Slag inclusion (schematic).
Source: MODENESI et al. 2011

2.4.6Lack of fusion

According to Wainer et al (2010), lack of fusion is when the joining of the filler material and the base metal does not completely fuse. It can happen due to a lack of heating of the metal present in the joint (lack of welding energy); an oxide layer (rust or scale) with a thickness that can complicate the fusion of the base metal; poor use of the torch, inadequate cleaning at the welding site or even in some places where the arc is unable to reach the appropriate areas of the joint. (MODENESI et al. 2009).

Lack of fusion is known as the stage in the process where the complete filling of the weld in the given area is not achieved. This can often be due to parameter errors, such as welding speed, too low a current, etc. (MODENESI et al. 2011).

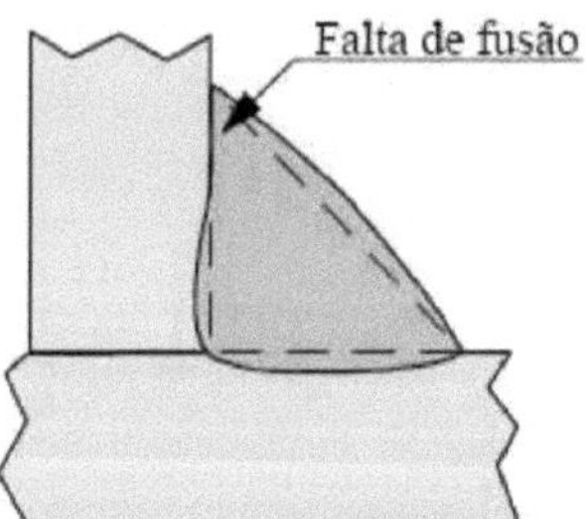

Figure 9 Lack of fusion (schematic) Source: (MODENESI et al. 2011).

"Lack of fusion is a discontinuity characterised by the non-coalescence of part of the weld bead, on the side of the bevel or between beads in multi-pass welding. " (NOVAIS, 2010).

According to Fortes (2005), this is a type of discontinuity that is easy to prevent, resulting from an inappropriate welding current or an inconvenient welding speed.

There is no repair procedure for this discontinuity, so it is necessary to remove the nonconforming weld bead and repeat the welding process.

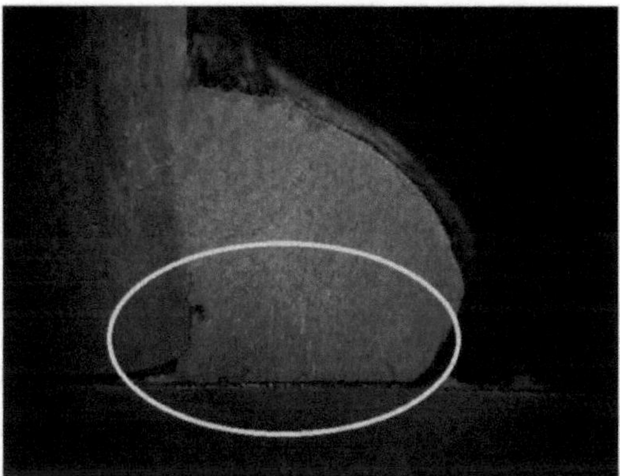

Figure 10 Lack of fusion in an angled joint Source: Alves and Maiorka (Unpublished article)

2. 5Ultrasound testing

According to Zolin (2011), sound is formed through the oscillation of a material, obtained in a high-pitched or low-pitched form, and at frequencies as low as 20 Hz, which are considered infrasound, or at frequencies above 20 kHz, which are called ultrasound.

According to Andreucci (2014), ultrasound is a type of non-destructive test capable of identifying defects or discontinuities on the inside of the area being analysed. These defects can originate from the manufacturing process of the elements to be analysed, such as: gas bubbles in castings, double lamination in rolled products, micro cracks in forgings, slag in welded joints and others.

Therefore, ultrasonic testing, like non-destructive testing, aims to reduce the amount of imprecision in the use of materials or liability parts (ANDREUCCI, 2014).

According to Silva; Mei (2010), verification using ultrasound equipment is a non-destructive process in which ultrasound waves (1-25 MHz) pass through the material being verified and thus provide assistance in finding any type of discontinuity that may exist.

> In the most classic form of the test, a sound wave pulse is introduced into a part surface, propagates through the material and the waves are reflected at interfaces. The reflected beam is detected and analysed to define the presence and location of flaws. A single transducer (head) is often used, capable of introducing an ultrasound pulse and detecting any reflected beam (SILVA; MEI, 2010, p. 589).

This method makes it easier to detect discontinuities such as cracks, flakes, pores, solidification cavities and metal-gas interfaces. During welding, elements of different kinds can be added, and these are also detected by ultrasound, as they cause partial reflections or scattering of the waves. As this process only shows wave reflections, it is not possible, on an industrial scale, to identify the specific type of defect that the wave reflection indicates (SILVA; MEI, 2010). Figure 11 below illustrates the basic response scheme obtained in ultrasound testing.

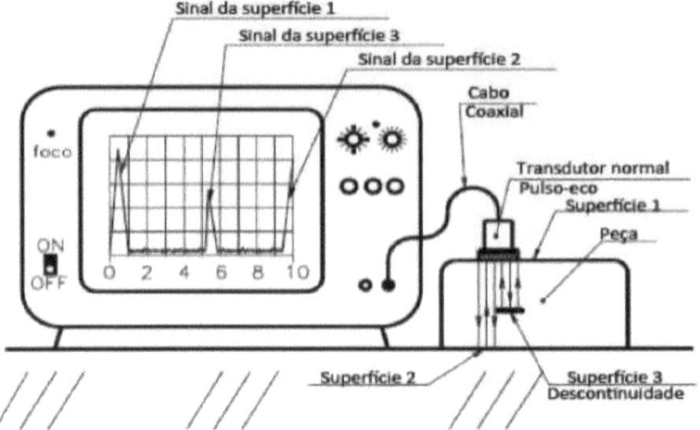

Figure 11 Basic schematic of the response obtained in the ultrasound test
Source: Unknown

According to Zolin (2011), in terms of the nature of waves, they are categorised as mechanical, as they require a material object to reproduce, for example the sound of a guitar string moving through the air, or electromagnetic waves, which do not require a material object to reproduce, for example X-rays and radio waves.

The speed at which these sound waves travel through materials depends on the conducting object and the direction of vibration (transverse and longitudinal waves). Table 2 shows the proliferation speed values for different material objects (ZOLIN, 2011).

Table 2 Speed of sound propagation

Speed of sound proliferation		
Material	Speed (m/s)	
	Longitudinal wave	Transverse wave
Aluminium	6300	3100
Lead	2160	700
Steel	5900	3250
Cast iron	3500 a 5600	2200 a 3200
Brass	3830	2050
Glass	5570	3520
Acrylic	2730	1430

Source: Zolin, 2011

Ultrasound tests use waves at frequencies between 0.5MHz and 25 MHz (500,000Hz to 25,000,000MHz) that are generated by a transducer. Piezoelectric crystals are objects that have the ability to transform mechanical pressure into electrical voltage and vice versa. In this situation, they transform electrical excitation energy into mechanical vibration energy while maintaining the electrical frequency, i.e. capturing mechanical energy and converting it into electrical energy

(ZOLIN, 2011).

Figure 12 shows how an ultrasound test works, which is summarised in the transmission of a mechanical wave by a transducer (a), from the moment the wave is transmitted the device starts counting time. The moment this wave hits a discontinuity, it is reflected back to the device (b) and reproduces an electrical signal, which is processed and then displayed on the screen of the analyser (d), making it possible to see the position of the echo formed by the path taken by the sound to the discontinuity in the material.

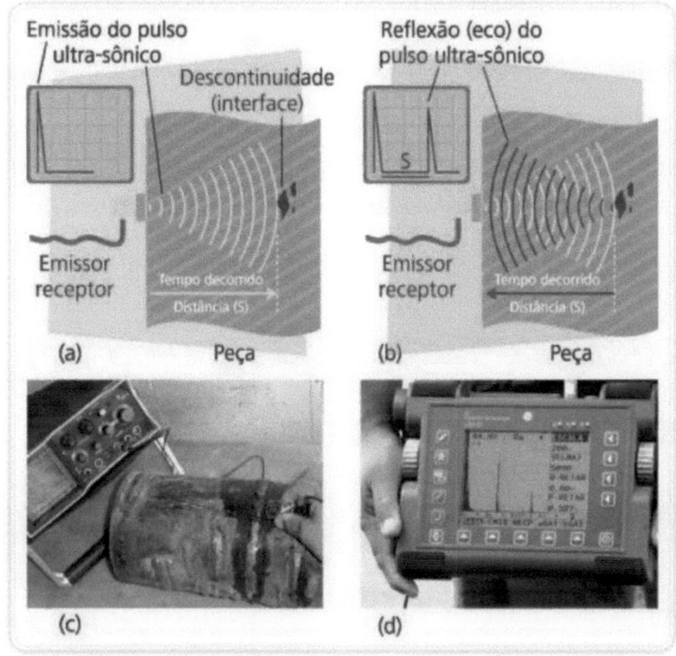

Figure 12 a) Transmission of the ultrasound pulse; b) echo generated by the reflection of the wave in the discontinuity; c) inspection of a part using ultrasound and d) detail of the graph formed by the ultrasound emission and echo.

Source: Zolin, 2011.

2. 6Welding positions   1G, 2G, 3G

1G means a weld made between metal plates in a flat position, 2G is another type of weld between plates, but made in a horizontal position and 3G is a type of weld between plates made in a vertical position, all shown in figure 13 below.

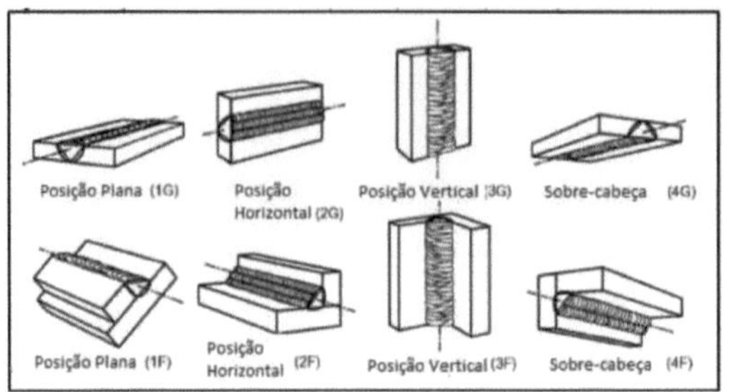

Figure 13 Welding positions, with chamfer (top) and fillet (bottom)

Source: GALVERY, 2006 apud BRAGNOLI, 2015

2. 7Ceramic backing

Ceramic backing is a product that has an aluminium adhesive foil and a ceramic material with a shape suited to the type of joint to be welded. Its application is simple and does not require specialised personnel. Figure 14 below shows a type of ceramic backing.

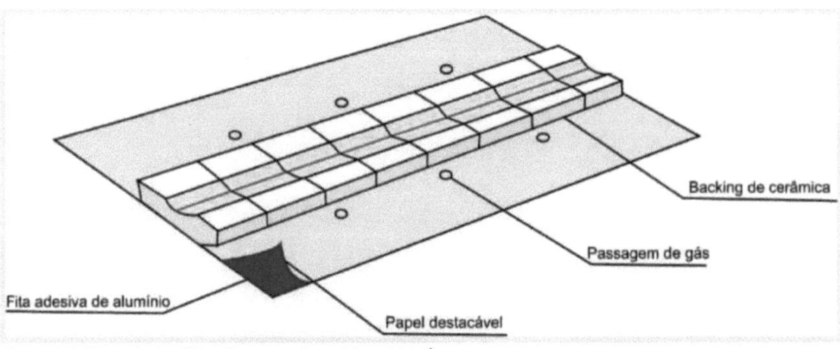

Figure 14 Ceramic backing Source: cig soldas, 2016[2]

2. 8Scientific Analytical Methods

It is a quantitative research method that requires a more detailed analysis of the information collected in a defined field of study, whether by observation or experiment, in a trial or test, with the intention of clarifying a circumstance of a certain event in a certain defined area. The real difference between an analytical study and a descriptive one is the analytical efficiency of making predictions for a population, from which samples are taken and statistical conclusions are drawn by testing hypotheses (FONTELLES, 2009).

---

[2] Ceramic backing. Cig soldas. Available at: < http://cigsoldas.com.br/backing-de-ceramica/> Accessed on: Nov. 2016.

CHAPTER 3

METHODOLOGY

3.1 Specimens (CP)

Specimens were produced to assess slag formation. They were made using ceramic backing, which can be used in the 1G, 2G and 3G positions. For this study, the specimens were produced in the 1G position.

The PCs were prepared as follows: two sheets measuring 150 mm wide, 200 mm long and 10 mm thick were cut and given a 60° bevel at one end, as shown in Figures 15, 16 and 17 below,

Figure 15. Preparation of the CPs
Source: Author, 2017

Figure 16 Cutting the steel sheets

Source: Author, 2017

Figure 17 60° chamfer

Source: Author, 2017

These CPs were then assembled, leaving a distance of 5 mm between one plate and the next, as shown in figure 18 below.

Figure 18 Distance between plates of 5 millimetres
Source: Author, 2017

At the bottom, two metal supports are fixed to ensure that the CP does not move while it is receiving the welding process, after which the ceramic backing is fixed to the bottom of the CP to help contain the deposition metal, as shown in figure 19 below.

Figura 19   Metal and ceramic backing
Source: Author, 2017

The PCs were subjected to a welding process known as MIG, as it was carried out with a mixture of CO_2 plus argon, using a gas flow rate in a closed environment of 15 litres per minute. The deposition wire is of the solid type with a diameter of 1.2 mm, as shown in figure 20 below, the voltage is between 23 and 31 (V) and the current is between 153 and 236 (A).

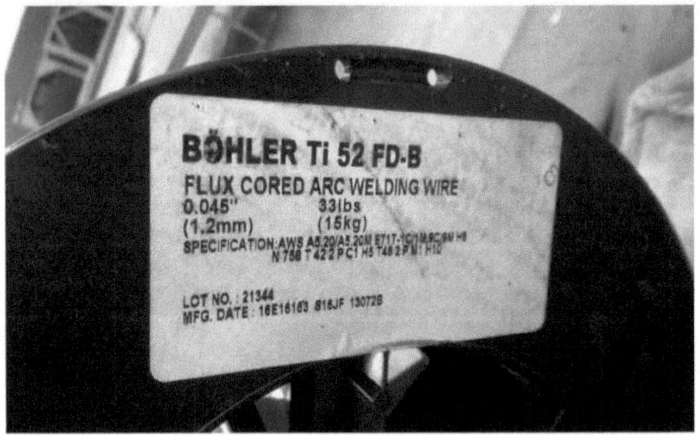

Figura 20   Deposition wire

Source: Author, 2017

These voltage and current values were for the processes carried out in the preparation of these CPs.

The CPs were analysed using ultrasound testing to assess the formation of total slag, as shown in Figures 21, 22 and 23 below.

Figure 21 Ultrasound test

Source: Author, 2017

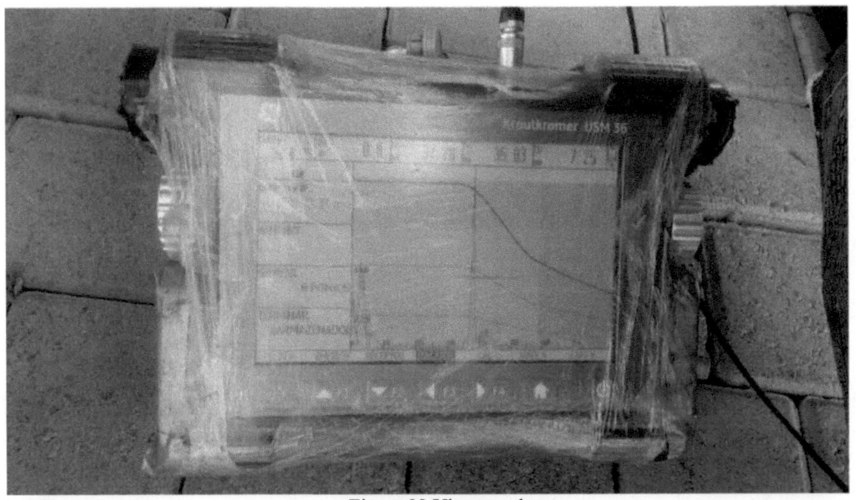

Figure 22 Ultrasound test

Source: Author, 2017

Figure 23 Ultrasound test Source: Author, 2017

3.2 Slag reduction tests   on the first strand

3.2.1 Optimising   the process parameters of the welding application

A specification of welding parameters was produced for the elaboration of a test specimen (CP). These process optimisation parameters were analysed in order to assess the occurrence of reduced slag formation in the first weld bead.

The details of the parameters are shown in table 3, where a number of CPs were manufactured for each angle, as specified in table 4 below:

Table 3 Specification of welding parameters.

Welding angle	75°	90°	105°
Gas flow (l/min)	15	25	35
Wire length (mm)	15	18	20

Source: Author, 2017

Table 4 Manufacture of CP.

Specification	Angle	Gas flow (l/min)	Wire length (mm)	Qty.
1°	75°	15	15	1
	75°	15	15	1
	75°	15	15	1
2°	90°	15	15	1
	90°	15	15	1
	90°	15	15	1
3°	105°	15	15	1
	105°	15	15	1
	105°	15	15	1

Source: Author, 2017.

After welding was completed, the PCs were subjected to non-destructive tests, which were carried out using ultrasound to assess the appearance of some non-conformities, such as: cracks, porosity, slag on the inside, lack of fusion. These tests also helped to check the quality of the welding and whether the parameters produced for the test matched the practice. In this case, the results were used to identify whether any discontinuities in the CPs were caused by the welding equipment, a lack of weld cleanliness or a lack of experience on the part of the employees carrying out the process.

CHAPTER 4

ANALYSING THE RESULTS

Three specimens were analysed for each item, following table 04 in the Methodology chapter. For the first specification, the results shown in figure 24 below were obtained.

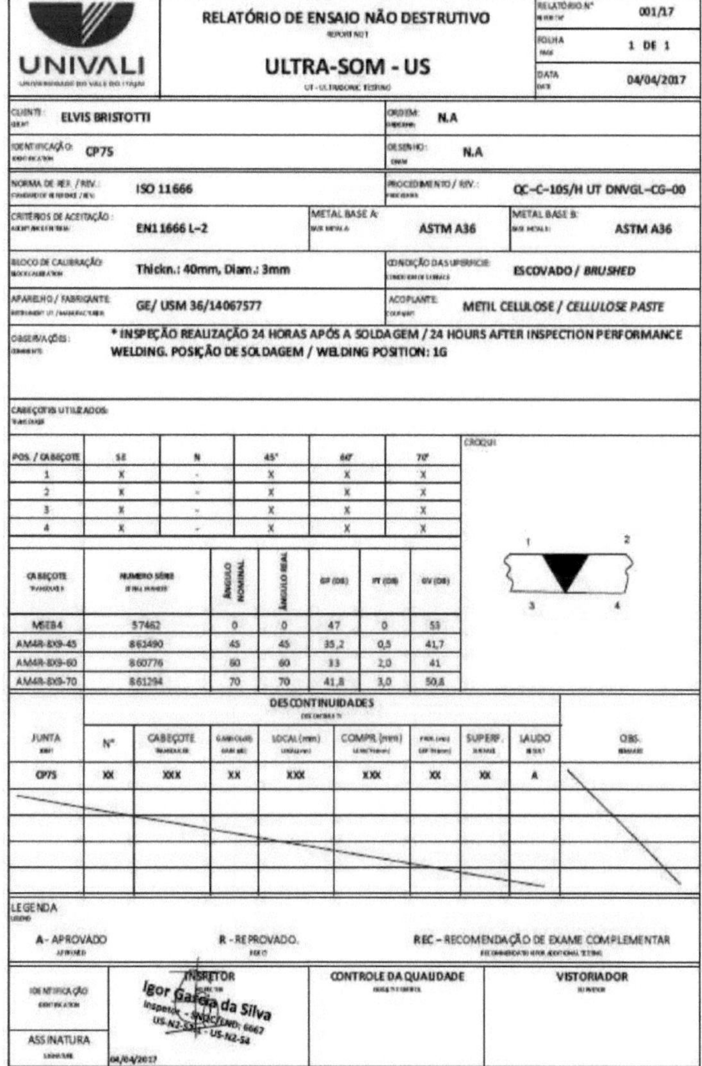

RELATÓRIO DE ENSAIO NÃO DESTRUTIVO (REPORT N°1)

ULTRA-SOM - US (UT - ULTRASONIC TESTING)

RELATÓRIO N°	001/17
FOLHA	1 DE 1
DATA	04/04/2017

CLIENTE: ELVIS BRISTOTTI		ORDEM: N.A	
IDENTIFICAÇÃO: CP75		DESENHO: N.A	
NORMA DE REF. / REV.: ISO 11666		PROCEDIMENTO / REV.: QC-C-105/H UT DNVGL-CG-00	
CRITÉRIOS DE ACEITAÇÃO: EN11666 L-2	METAL BASE A: ASTM A36	METAL BASE B: ASTM A36	
BLOCO DE CALIBRAÇÃO: Thickn.: 40mm, Diam.: 3mm		CONDIÇÃO DAS SUPERFÍCIE: ESCOVADO / BRUSHED	
APARELHO / FABRICANTE: GE/ USM 36/14067577		ACOPLANTE: METIL CELULOSE / CELLULOSE PASTE	

OBSERVAÇÕES: * INSPEÇÃO REALIZAÇÃO 24 HORAS APÓS A SOLDAGEM / 24 HOURS AFTER INSPECTION PERFORMANCE WELDING. POSIÇÃO DE SOLDAGEM / WELDING POSITION: 1G

CABEÇOTES UTILIZADOS:

POS. / CABEÇOTE	SE	N	45°	60°	70°
1	X	-	X	X	X
2	X	-	X	X	X
3	X	-	X	X	X
4	X	-	X	X	X

CABEÇOTE	NÚMERO SÉRIE	ÂNGULO NOMINAL	ÂNGULO REAL	GP (DB)	PT (DB)	GV (DB)
MSEB4	57462	0	0	47	0	53
AM4R-8X9-45	861490	45	45	35,2	0,5	41,7
AM4R-8X9-60	860776	60	60	33	2,0	41
AM4R-8X9-70	861294	70	70	41,8	3,0	50,8

DESCONTINUIDADES

JUNTA	N°	CABEÇOTE	GANHO (dB)	LOCAL (mm)	COMPR. (mm)	PROF. (mm)	SUPERF.	LAUDO	OBS.
CP75	XX	XXX	XX	XXX	XXX	XX	XX	A	

LEGENDA

A- APROVADO   R - REPROVADO.   REC – RECOMENDAÇÃO DE EXAME COMPLEMENTAR

IDENTIFICAÇÃO	INSPETOR: Igor Garcia da Silva Inspetor - US N2 / END: 6662 US-N2-SC - US-N2-54	CONTROLE DA QUALIDADE	VISTORIADOR
ASSINATURA	04/04/2017		

Figure 24 First specification report

Source: Author, 2017

For the welding conditions, configured with a welding angle of 75°, a gas flow rate of 15 (l/min) and a wire length of 15 (mm), the ultrasound analysis proved that the process had met the optimum welding specifications.For the second specification, the results shown in figures 25 and 26 below were obtained.

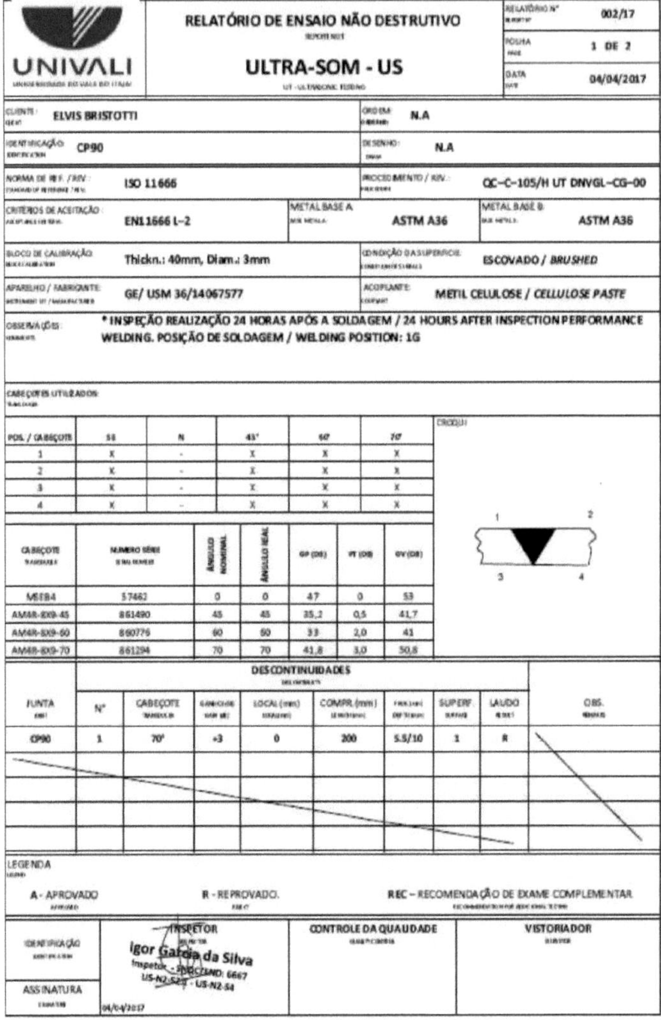

UNIVALI	RELATÓRIO DE ENSAIO NÃO DESTRUTIVO REPORT N°1	RELATÓRIO N° REPORT N°	002/17
		FOLHA PAGE	1 DE 2
UNIVERSIDADE DO VALE DO ITAJAÍ	ULTRA-SOM - US UT - ULTRASONIC TESTING	DATA DATE	04/04/2017

CLIENTE: CLIENT	ELVIS BRISTOTTI	ORDEM: ORDER	N.A
IDENTIFICAÇÃO IDENTIFICATION	CP90	DESENHO: DRAW	N.A
NORMA DE REF. / REV.: STANDARD OF REFERENCE / REV.	ISO 11666	PROCEDIMENTO / REV.: PROCEDURE	QC–C–105/H UT DNVGL–CG–00

CRITÉRIOS DE ACEITAÇÃO: ACCEPTANCE CRITERIA	EN11666 L–2	METAL BASE A BASE METAL A: ASTM A36	METAL BASE B BASE METAL B: ASTM A36
BLOCO DE CALIBRAÇÃO: BLOCK CALIBRATION	Thickn.: 40mm, Diam.: 3mm	CONDIÇÃO DA SUPERFÍCIE: CONDITION OF SURFACE	ESCOVADO / BRUSHED
APARELHO / FABRICANTE: INSTRUMENT UT / MANUFACTURER	GE/ USM 36/14067577	ACOPLANTE: COUPLANT	METIL CELULOSE / CELLULOSE PASTE

OBSERVAÇÕES:
REMARKS:
* INSPEÇÃO REALIZAÇÃO 24 HORAS APÓS A SOLDAGEM / 24 HOURS AFTER INSPECTION PERFORMANCE WELDING. POSIÇÃO DE SOLDAGEM / WELDING POSITION: 1G

CABEÇOTES UTILIZADOS:
TRANSDUCER

POS. / CABEÇOTE	SE	N	45°	60°	70°
1	X	-	X	X	X
2	X	-	X	X	X
3	X	-	X	X	X
4	X	-	X	X	X

CABEÇOTE TRANSDUCER	NÚMERO SÉRIE SERIAL NUMBER	ÂNGULO NOMINAL	ÂNGULO REAL	GP (DB)	PT (DB)	GV (DB)
MSEB4	57462	0	0	47	0	53
AM4R-8X9-45	861490	45	45	35,2	0,5	41,7
AM4R-8X9-60	860776	60	60	33	2,0	41
AM4R-8X9-70	861294	70	70	41,8	3,0	50,8

DESCONTINUIDADES
DISCONTINUITY

JUNTA JOINT	N°	CABEÇOTE TRANSDUCER	GANHO+DB GAIN+dB	LOCAL (mm) LOCATION	COMPR. (mm) LENGTH(mm)	PROF. (mm) DEPTH(mm)	SUPERF. SURFACE	LAUDO RESULT	OBS. REMARKS
CP90	1	70°	+3	0	200	5.5/10	1	R	

LEGENDA
LEGEND

A - APROVADO
APPROVED   R - REPROVADO.
FAIL   REC – RECOMENDAÇÃO DE EXAME COMPLEMENTAR
RECOMMENDATION FOR ADDITIONAL TESTING

IDENTIFICAÇÃO IDENTIFICATION	INSPETOR INSPECTOR Igor Garcia da Silva Inspetor - PROCIND: 6667 US-N2-S24 - US-N2-S4	CONTROLE DA QUALIDADE QUALITY CONTROL	VISTORIADOR SURVEYOR
ASSINATURA SIGNATURE	04/04/2017		

Figure 25 Second specification report

Source: Author, 2017

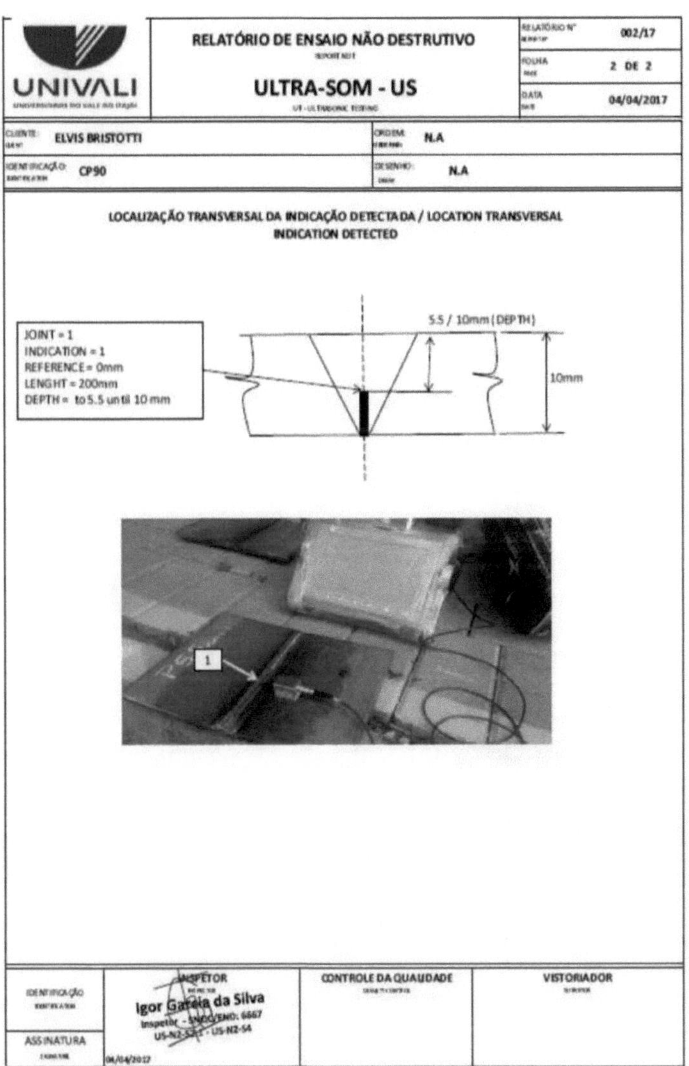

Figure 26 Second specification report

Source: Author, 2017

For the welding process parameters set at a welding angle of 90°, gas flow rate of 15 (l/min) and wire length of 15 (mm), the above report proves that the specimens were rejected due to lack of penetration.For the third specification, the results shown in figures 27 and 28 below were obtained.

Figure 27 Third specification report

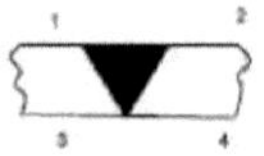

Source: Author, 2017

Figure 28 Third specification report

Source: Author, 2017

For the welding process parameters set at a welding angle of 105°, gas flow rate of 15 (l/min) and wire length of 15 (mm), the above report proves that the specimens were rejected due to a lack of penetration.

According to the welder responsible for making the welds on the test specimens, the position has a great influence on the time of execution, making it difficult to achieve the excellence required to reach the optimum point of the proposed specification, as we can see in figure 29 below that the welder is totally without an adequate position to carry out the process and this leads to the discontinuity that was presented in the ultrasound report.

Figure 29 Welding position Source: Author, 2017

CHAPTER 5

FINAL CONSIDERATIONS

It can be concluded from this work that the CPs with an angle of 75° reached the optimum level of the welding process, since after the ultrasound test none of the types of discontinuity were identified.

The 90° and 105° welding angles may not be viable for the process, due to the lack of penetration. Logically, this observation would need to be confirmed with more test specimens.

It was not possible to carry out the tests on all the gas flow and wire length specifications provided for in the previous methodology presented at the beginning of this work, because the company responsible for supporting the research worked with a gas network at a constant standard flow rate of 15 litres per minute, and would not agree to vary this for process reasons.

As for the length of the wire, because it is measured in millimetres, the welder was unable to keep to the standard laid down by the methodology, because the movement involved in carrying out the welding process directly influences compliance with the height of the wire (ergonomic issues).

It would be valuable for this work if the parameters of the research process stipulated in its methodology were developed in future works to complement the knowledge.

During the course of this research, some problems stood out, such as the delay in obtaining authorisation to make the test specimens, adapting the company's existing process to the specifications proposed in this work and not being able to carry out all the proposed specifications due to a lack of resources.

CHAPTER 6

REFERENCES

ALVES, D.; MAIORKA, J. V. **Macrographic analysis and hardness measurements of the heat-affected zone of a welded joint**. University of Passo Fundo. Passo Fundo, 6 p. Unpublished paper.

ANDREUCCI, R. **Industrial application of ultrasound testing.** 07/ 2016. Available at: <http://www.abendi.org.br/abendi/Upload/file/biblioteca/apostila_us_2016.pdf> Accessed on: 06/09/2016.

BRAZIL. **Law 3.381, of 24 April 1958**. Federal Official Gazette - Section 1 - 25/4/1958, Page 9385.

BROGNOLI, A. L. **Automation in industrial ship welding. Comparison of the mig system by welding trolley and manual processes**. 2016. 80 f. Scientific and Technological Initiation Work (Shipbuilding Technologist) - Universidade do Vale do Itajaí, Itajaí, 2016. [Supervisor: Prof Jucimar J. Aninha].

CIRIBELLI, M. C. **Como Elaborar uma Dissertação de Mestrado Através da Pesquisa Científica**. Rio de Janeiro: 7Letras, 2003.

DOS SANTOS, G. S. **Analysing the evolution of the naval industry**. 2011. 73 f. Scientific and Technological Initiation Work (Technologist in Shipbuilding) - Universidade Estadual da Zona Oeste, Rio de Janeiro, 2011. [Supervisor: Prof Érico Vinícius Haller dos Santos da Silva].

FILHO, E. G. M. **Study of the relative influence of mig-mag (gmaw) welding process variables on the root pass**. 2012. 91 f. Master's dissertation (Master's in Materials Engineering) - Federal Centre for Technological Education of Minas Gerais, Belo Horizonte, 2012. [Supervisor: Profa Maria Celeste Monteiro de Souza Costa, Dr].

FONTELLES, M. J; SIMÕES, M. G; FARIAS, S. H; FONTELLES, R. G. S. **Metodologia da pesquisa científica: diretrizes para a elaboração de um protocolo de pesquisa,** Belém - Pará. Aug. 2009.

FORTES, C. ESAB: **Welding metallurgy workbook**. Contagem: [s.n], 2005. Available at: <http://www.esab.com.br/br/pt/education/apostilas /upload/apostilametalurgiasoldagem.pdf>. Accessed on 24 July 2016.

FORTES, C. ESAB: **Tubular Wire Workbook**. Contagem: [s.n], 2006. Available at: <http://www.esab.com.br/br/pt/education/apostilas/upload/1901098rev1_apostilaaramestubul ares_ok.pdf>. Accessed on 06 Sep. 2016.

HORMEU, H. **Module II: Basic Welding.** Electro-electronics Technical Course. July 2008. Federal Technological Education Centre of Santa Catarina Araranguá Unit.

JUNIOR, L. P. D. S., CABRAL, T.D.S. **Specification of mig welding procedures for filling cavities using the successive layer technique.** 2008. 78 f. Scientific and Technological Initiation Work (Bachelor's Degree in Mechanical Engineering) - Federal University of Pará, Belém, 2008. [Supervisor: Prof. Eduardo de Magalhães Braga].

MACHADO, E. **Influence of winds on the quality of welds carried out in shipyards using the flux-cored wire process.** 2015. 67 f. Scientific and Technological Initiation Work (Bachelor in Naval Engineering) - Federal University of Santa Catarina, Joinville, 2015. [Supervisor: Prof. Dr Tiago Vieira da Cunha].

MACHADO, I. G. **Welding and related techniques: processes.** Porto Alegre, 1996.

MODENESI, P. J.; MARQUES, P. V.; BRACARENSE, A. Q. **Soldagem: Fundamentos e Tecnologia**. 3. ed. BELO HORIZONTE: UFMG, 2009.

MODENESI, P. J.; MARQUES, P. V.; BRACARENSE, A. Q. **Soldagem: Fundamentos e tecnologia**. 3ª updated edition- Belo Horizonte: UFMG, 2011.

NOVAIS, P. R. S. **Evaluation of the main discontinuities found in welded joints, causes and possible solutions.** Available at: <http://www.abcem.org.br/construmetal/2010/downloads/contribuicoes-tecnicas/ct09.pdf> Accessed on: 06/09/2016.

QUITES, A. M. **Introdução à soldagem a arco voltaico**.ia edição - Florianópolis: SOLDASOFT, 2002.

SANTO, A. do E. **Delineamentos de metodologia Científica**. São Paulo, Loyola editions, 1992.

SITE METALICA.COM.BR. **Characterisation of a welded joint with a glass fibre-based welding support.** Available at: <http://wwwo.metalica.com.br/caracterizacao-de- uma-junta-soldada-com-suporte-de-solda-a-base-de-fibra-de-vidro> Accessed on: 16/08/2016.

SITE INFOSOLDA.COM.BR. **Discontinuities.** Available at:

<http://www.infosolda.com.br/biblioteca-digital/28-biblioteca-digital/livros- senai/metalurgia/125-descontinuidades.html> Accessed on: 13/09/2016.

CIG **ceramic backing** website. Available at: <http://cigsoldas.com.br/backing-de- ceramica/> Accessed on: 27/11/2016.

SILVA, A. L. C.; MEI, P. R. **Special steels and alloys.** 3ª edition - São Paulo: Blucher, 2010.

WAINER, E.; BRANDI, S. D.; MELLO, F. D. H. de. **Welding**: Processes and Metallurgy. São Paulo: Edgard Blucher, 1992.

WAINER, E.; BRANDI, S. D.; de MELLO, F. D. H. **Soldagem, Processos e Metalurgia**. São Paulo, Edgar Blucher. 2010.

ZOLIN, I. **Technical course in industrial automation: mechanical tests and failure analyses** / Ivan Zolin. - 3. ed. - Santa Maria: Federal University of Santa Maria: Industrial Technical College of Santa Maria, 2010. 102 p.: ill.

Printed by Books on Demand GmbH, Norderstedt / Germany